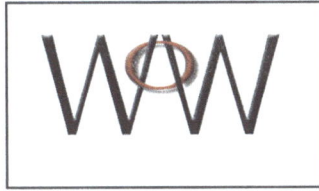

Choosing a
Career in Mortuary
Science and the
Funeral Industry

Medical examiners carry the body of an unidentified woman found on a beach into a morgue for an autopsy.

Choosing a Career in Mortuary Science and the Funeral Industry

Nancy L. Stair

The Rosen Publishing Group, Inc.
New York

To GK and Donna (M/D),
whose last trip will be to
Rocky Mountain National Park
in Colorado

Published in 2002 by The Rosen Publishing Group, Inc.
29 East 21st Street, New York, NY 10010

Library of Congress Cataloging-in-Publication Data

Stair, Nancy L.
Choosing a career in mortuary science and the funeral industry / Nancy L. Stair.
p. cm. — (World of work)
Includes bibliographical references and index.
Summary: Explores career opportunities for those interested in becoming a funeral director, embalmer, coroner, medical examiner, or pathologist.
ISBN: 978-1-4358-8856-2
1. Forensic pathologists—Vocational guidance—Juvenile literature. 2. Coroners—Vocational guidance—Juvenile literature. 3. Undertakers and undertaking—Vocational guidance—Juvenile literature. [1. Forensic pathologists—Vocational guidance. 2. Coroners—Vocational guidance. 3. Undertakers and undertaking—Vocational guidance. 4. Vocational guidance.]
I. Title. II. World of work (New York, N.Y.)
RA1063.4 .S73 2001
363.7'5'02373—dc21

2001002497

Manufactured in the United States of America

Contents

Introduction

You've probably heard the old saying, "Only two things in life are certain: death and taxes." It's true: At some point, we all will die. Although you may think death is an unpleasant subject, it is something that everybody must cope with. People are different all over the world, but death is one thing that everyone has in common.

Have you ever wondered what happens to a dead body? Who takes care of it before the funeral? What does the family need to do? When someone dies, the survivors may feel guilty and confused. They are too distraught to cope with all of the details of taking care of the body.

Fortunately, there are trained professionals to help friends and family of the deceased through this distressing time. Funeral industry professionals pick up and prepare the body, arrange the funeral, and help the surviving family and friends honor the deceased. It may seem like a morbid job to do, but it's actually an important service to the living as well as the dead.

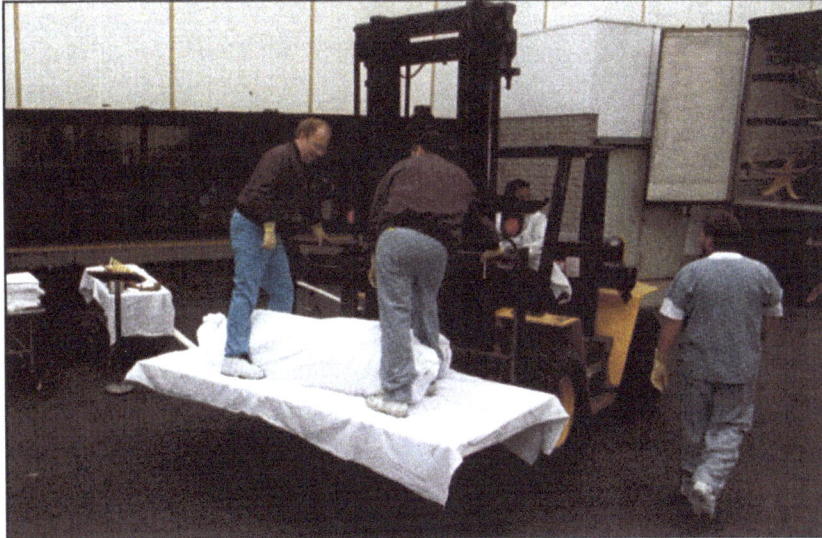

One of the tasks of working in a medical examiner's office is moving bodies from the scenes of accidents, murders, and suicides. It can be a grim job, but these workers do it with respect and dignity for the deceased.

For those who work with death on a daily basis, it's often more than just a job. "It's almost a calling," says Denise R. Hammelrath, a funeral director and embalmer who teaches at Cincinnati College of Mortuary Science in Cincinnati, Ohio. "It's a caring profession, but with the best of both worlds: On the one hand, it's science, and on the other hand, it's sociology. Plus, you're not doing the same thing all day long. It's never boring."

There are many different jobs in the funeral industry. If you choose a career as a medical examiner (ME) or coroner, you will go to murder and accident scenes and use your knowledge of science to investigate crimes. As a pathologist, you will examine bodies inside and out. As a funeral director, you may be called out of bed in the middle of the night to start preparations for embalming. No matter what you do, you will take care of those who have suffered the inevitable.

1

Investigating Death

Just as no two lives are the same, every death is unique. Funeral industry professionals must be prepared to perform many different duties when the phone rings and someone's demise is reported.

What Happens After a Person Dies?

Funeral industry professionals spring into action after a person dies. If a person dies suddenly, as in an accident, others may call an ambulance, the police, or 911. When an ambulance arrives, the emergency medical technicians (EMTs) or paramedics respond. Paramedics usually are required by law to try to revive anyone who is not clearly dead. Once they are certain that a person is dead, they take the body to the hospital morgue. If a death occurs in the hospital, hospital orderlies take the body to the morgue. If a person dies at home, a funeral director takes the body to a funeral home. If foul play is suspected, the medical examiner or coroner will pick up the body and take it to the medical examiner's office or to the morgue.

If someone dies while traveling, things can get more complicated. In such cases that occur in the United States and Canada, local hospitals can refer next of kin to funeral homes for embalming (embalming is discussed in detail in chapter 3), which is required by law before a body can cross state lines. If a person dies outside of the United States, the body must be embalmed before it can be brought into the country.

In most countries, including the United States and Canada, laws require that a doctor must pronounce a person dead. If a person dies outside of the hospital, the body usually must be taken to a hospital to have a doctor make a formal pronouncement of death. After a doctor or other professional makes this pronouncement, he or she must sign a death certificate. The death certificate should list the "underlying causes" (such as disease or injury) which led directly to death, or the circumstances of the injury that caused the death (a traffic accident or a shooting, for instance). The doctor sends the death certificate to the funeral director, to local and state government offices, and to federal government bureaus that study the causes of death. This is true in both the United States and Canada.

The death certificate is important for the family of the deceased. It is required to obtain permission to bury the deceased, make life insurance claims, receive death benefits from the government, and settle credit card bills or debts.

Sometimes an autopsy must be performed to determine the cause of death. For instance, it may

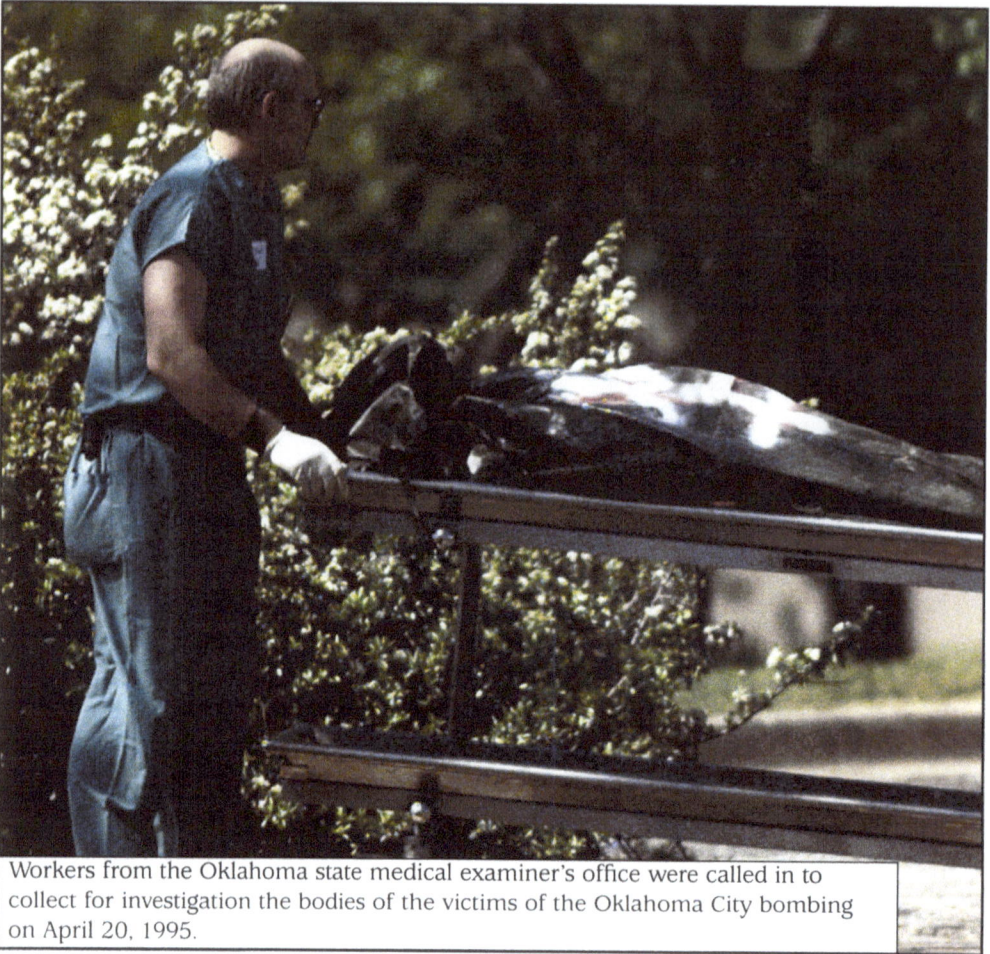

Workers from the Oklahoma state medical examiner's office were called in to collect for investigation the bodies of the victims of the Oklahoma City bombing on April 20, 1995.

appear as if a man died in a car crash, only for an autopsy to later reveal that he was shot while driving, causing him to crash his car. So instead of an accident, it was a murder. According to laws in the United States and Canada, if a death is unexpected, suspicious, violent, work-related, thought to be caused by a communicable disease, has occurred in prison, or will result in cremation, the case must be referred to the medical examiner or coroner.

If an autopsy is necessary, the police, a doctor, or a funeral director must call the medical examiner or coroner. Medical examiners and coroners perform two different jobs, but they provide essentially the same service to their communities: They investigate deaths. In the United States, state laws specify the types of deaths that must be investigated and who— either a coroner or a medical examiner—should do the job. Often, the terms are used interchangeably. All states in the United States have formal death investigation systems that use medical examiners, coroners, or a combination of the two. According to United States government statistics, twenty states have medical examiners, eleven use coroners, and the remaining states use mixed systems.

What Is a Coroner?

A coroner is an appointed or elected public official who directs the investigation of deaths. The job dates back to medieval days in England, when the chief purpose of a "crowner" was to make sure that death taxes were paid to the king. Today, a coroner's duties are much more challenging. He or she must determine whether a death occurred under natural circumstances or if it was due to accident, homicide, or suicide. Coroners must investigate every death thoroughly to assign each a specific cause and manner. This information appears on the official death certificate.

Depending upon where you live in the United States or Canada, training to become a coroner varies. In some regions, the coroner must be a physician and perform autopsies. But in many locations, the coroner is not required to be a doctor or even trained in medicine. Coroners who live in regions that do not require a medical degree usually hire physicians, pathologists, or forensic

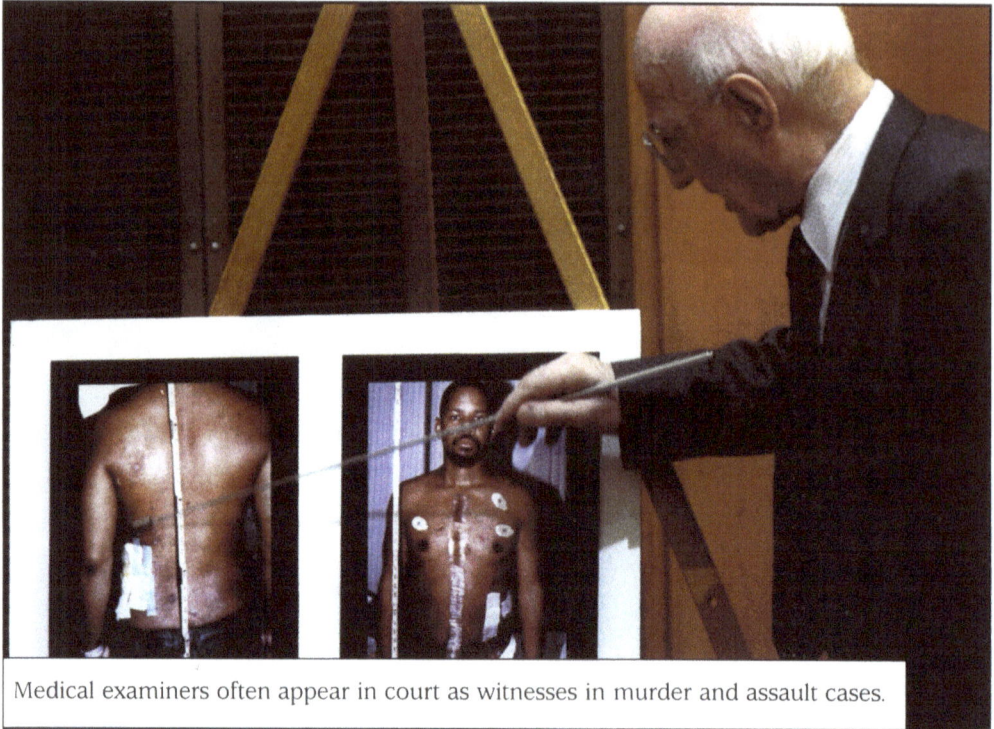

Medical examiners often appear in court as witnesses in murder and assault cases.

pathologists to perform autopsies. Those who live in areas that require coroners to possess a degree in medicine need to follow the career path and education of the medical examiner.

What Is a Medical Examiner?

The medical examiner (ME) position is an American creation that has existed only for a century. Medical examiners are almost always appointed to their positions, and must be licensed physicians with extensive formal training in medical and legal death investigations. Unlike a coroner, a medical examiner is expected to use his or her medical expertise to find out what killed a person.

Medical examiners often visit the scenes of deaths or crimes to examine corpses and look for evidence that the police may not recognize as being related to the cause of death. They need to determine the identity of the deceased person, the exact time of death, the manner of death (natural, disease, accident, suicide, or homicide), and the medical cause of death. They want to know if a weapon or other object caused the death. Medical examiners often collect evidence, such as bullets, hair, fibers, bodily fluids, and trace chemicals, in order to reconstruct the way a person died.

If you want to become a medical examiner, you must earn a degree in medicine. There are many different specialties involved in a death investigation, but a medical examiner usually is not required to know everything. Medical examiners often hire forensic scientists to perform autopsies in order to determine the cause of a person's death, and to assist with different techniques necessary to conduct a precise and accurate investigation.

What Does a Pathologist Do?

Pathology is the study of disease and its causes. A pathologist is a doctor trained in the medical specialty of pathology. A forensic pathologist, or forensic scientist, is a doctor who specializes in the investigation of sudden, unexpected, and violent deaths. Often the terms are used interchangeably since these jobs involve the same kinds of skills and responsibilities.

Pathologists investigate deaths reported to a coroner or a medical examiner. They often go to death or crime scenes to examine dead bodies and look for drugs or toxins (poisons) that may be related to the cause of death. Like MEs, pathologists attempt to determine the identification of the deceased, time of death, manner of death, cause of death, whether the death was by injury, and the weapon or other object that may have caused the death. They also collect medical evidence, such as trace chemicals or secretions (fluids) on the body, in order to reconstruct how a person received the injuries that killed him or her. Pathologists investigate more thoroughly and more scientifically than MEs, who often act as supervisors to people performing the examinations.

Traditional medicine is not the only area of knowledge a pathologist needs to have. Sometimes, it is not clear how a person died. Since many different things can kill a person, a broad range of knowledge, education, and training is needed to solve the puzzle. For example, many people die of gunshots, so a pathologist must have knowledge of firearms to determine what kind of gun and ammunition were used in a murder. This information often helps police track down the gun and leads them to the killer.

Since some deaths are caused by poisonings, both accidental and intentional, a pathologist must be educated in toxicology, which is the study of poisons. Some poisons are not detected on a routine drug screen, so the pathologist must understand medications and toxins in order to request the specific analytical tests needed to detect them.

In the examination of "skeletonized" or severely decomposed human remains, a pathologist needs formal training in and a working knowledge of many kinds of identification procedures, such as the study

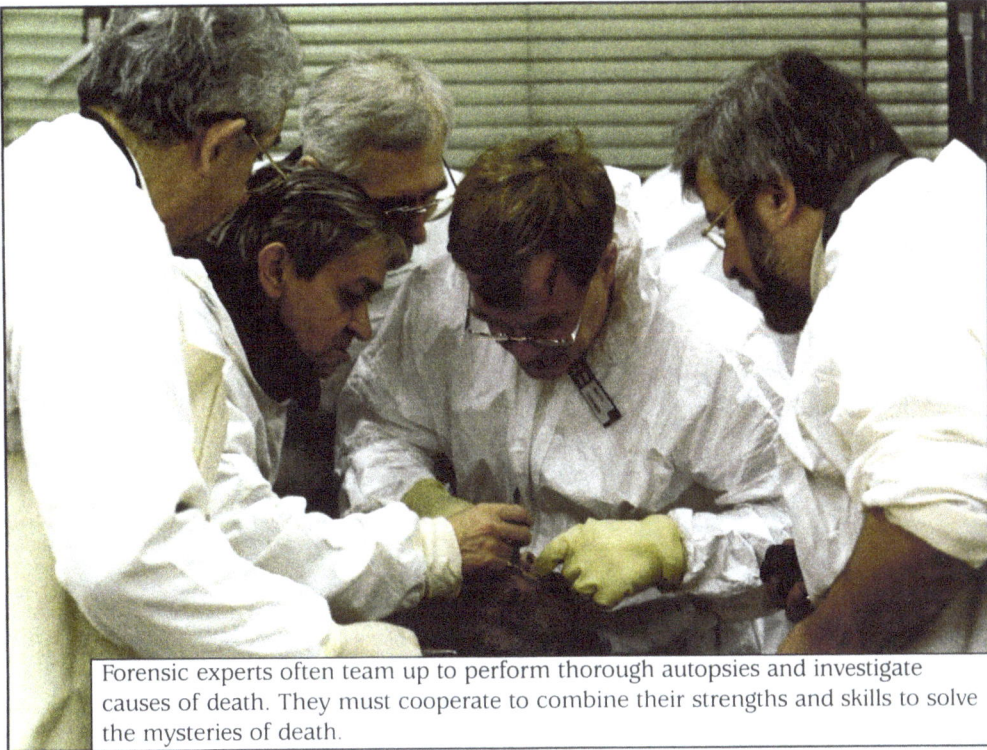

Forensic experts often team up to perform thorough autopsies and investigate causes of death. They must cooperate to combine their strengths and skills to solve the mysteries of death.

of bones and decomposition. He or she must be skilled in many techniques in order to establish a corpse's identity. Nowadays, forensic pathologists must often do DNA testing, which despite its growing use is still a complex procedure.

A medical examiner usually begins an investigation by gathering information from the police and witnesses about how they think the death in question occurred. A pathologist talks to the deceased's relatives or friends to try to put together a past medical history of the deceased. This information also may offer clues. Next, the pathologist examines the dead body for visible signs of the cause of death, such as bruises, gunshot wounds, knife wounds, or other injuries. He or she then must examine the body for internal clues to the manner of death by performing an autopsy.

Performing an Autopsy

Autopsies can teach us a lot about how a person died, and also about how he or she lived. There are many reasons for performing autopsies, but one of the most important is that the knowledge gained from studying the dead often helps the living.

Autopsies are also performed to identify new diseases that could be infectious, or spread easily between people. They can determine whether many people are dying of the same disease. If someone dies during surgery, an autopsy can explain why the medical procedure was unsuccessful, and reveal how it can be improved. Autopsies also are used to teach new medical students and to provide organs for research.

A hospital autopsy is often performed on individuals who have died from a known disease. The purpose of the autopsy is to determine the extent of the disease, the effects of treatment, and the presence of any undiagnosed malady that might have contributed to death. The next of kin must give permission for the autopsy and may limit the extent of the dissection (for example, allowing an autopsy of the chest and abdomen only, excluding the head).

A forensic autopsy that is ordered by the coroner or medical examiner is considered authorized by law and is performed in a morgue. (It is sometimes called a medicolegal autopsy.) During the course of the forensic autopsy, blood and other body fluids are routinely drawn in order to check for the presence of alcohol and other drugs. The forensic autopsy should encompass the entire body. Unlike a hospital autopsy, the next of kin does not authorize and may not limit the extent of a forensic autopsy.

In performing an autopsy, a pathologist cuts open the body of the deceased to check the organs and tissues for evidence of infection. He or she tests body fluids, cells, and tissues, examining many samples under a microscope to detect clues that are not visible to the naked eye. A solid scientific education (especially chemistry, biology, and anatomy) is essential, including experience in a laboratory.

When all of the information, including the history, the results of the autopsy, and the laboratory tests, is completed, the forensic pathologist puts all of the information together and draws conclusions about the cause and manner of death. He or she prepares a report summarizing the findings and can expect to be subpoenaed, or called to testify before courts about how people died. Forensic scientists are often involved in many aspects of criminal cases, and the results of their work may help solve crimes.

Coroners, medical examiners, and pathologists also provide copies of their official reports to those who have legitimate business and financial interests in the cause and manner of death, such as insurers or public health agencies. Autopsy data provided by coroners, medical examiners, and pathologists is studied by researchers and government health and safety agencies to learn how to prevent disease and injury, with the goal of saving lives in the future. For example, the data developed from studying car accident victims led to legislation requiring seat belts, and information about fire deaths is responsible for the widespread use of smoke detectors.

It can be important to conduct autopsies even when the cause of death appears obvious. In cases of individuals who have undergone medical treatment, it is important to share the findings with other doctors for their education. In cases of shootings or assaults, the forensic pathologist may recover bullets or other important evidence to catch criminals. In cases of car accidents, it is important to determine who was driving or if the vehicle malfunctioned, factors that might have caused or contributed to the crash. Forensic autopsies may identify inherited diseases, such as heart disease or kidney disease, that could be a risk for the deceased's next of kin.

Through modern medicine and investigative techniques, medical examiners, coroners, and pathologists use lessons learned from the dead to enhance the quality of life for the living. But there is more to the funeral industry than just science. When a loved one dies, grief and responsibilities can overwhelm the survivors. Fortunately, there are professionals who address this aspect of death.

2

The Funeral Industry

In most cultures, it was once traditional for family members to bury their own relatives when they died. Undertakers were primarily carpenters who built coffins on the side. But thanks to the affordability and convenience of transportation, and the necessity of immigration, family members began to live far apart from one another. Furthermore, stronger public health laws required that trained professionals dispose of the dead to prevent the spread of diseases. Eventually, tradition died in the face of science; in the United States and Canada, the handling and burial of the dead shifted entirely away from the deceased's family and friends, and now rests squarely upon the shoulders of the funeral director.

What Purpose Does a Funeral Serve?

The funeral is very important for the survivors of the deceased. For those who are left behind, a funeral provides a place for family and friends to gather for support and to reminisce. A funeral, memorial service, or other gathering permits family and friends to express their feelings, to offer comfort to one another, and to begin to accept the reality of loss. Friends and family of the deceased use this time to celebrate the life and accomplishments of the person they knew and loved. It's a chance for everyone who knew the deceased to say good-bye.

Just as people of all faiths have traditions and ceremonies for birth and marriage, they also have traditions and ceremonies for death and grief. A funeral service can be as unique as the individual being honored. Most often, a funeral includes a visitation or viewing period, also known as a wake. A viewing period is a time for family and friends to gather with the body of the deceased, allowing the living to confront the fact of death. In some faiths, it is necessary for family and friends to confirm that a life has been lived and is now over. After the wake, there is often a religious service, and then a burial or a cremation. In some cultures, an additional memorial service is held after a specific amount of time has passed. It is the job of the funeral director to arrange and direct all of these tasks for grieving families, and to relieve them of the work so they can grieve and say farewell to their loved one.

The Funeral Director

Funeral directors also are called morticians or undertakers. Funeral directors have been stereotyped in the past as scary or morbid people, but in reality they are hard-working, compassionate professionals who have a difficult job to do. This career may not appeal to everyone, but those who work as funeral directors take great pride in providing efficient and sensitive services for the family and friends of the deceased.

A funeral director arranges and handles all of the details of funerals. Family members must be interviewed so the director can learn what they want for the funeral, how they prefer the body to be treated, and which clergy or other persons will officiate. Together with the family, the funeral director establishes the location, dates, and times of wakes, memorial services, and burials. Sometimes the deceased leave detailed instructions for their own funerals, and the funeral director carries these out.

Funeral directors also arrange for pallbearers and clergy, schedule the digging and filling of the grave in the cemetery, decorate and prepare the sites of all services, and provide transportation between the service and the grave site for the remains, mourners, and flowers. Funeral directors usually rent or lease a hearse to carry the body to the funeral home or mortuary. If the deceased person is going to be buried in another state or country, the funeral director will have to prepare the remains and casket for shipping.

Funeral directors often work long, irregular hours. They work on-call, meaning that they carry a pager or a cell phone (probably both) and must take any call, no matter what time it is. This is because they may be needed to move a body from a home, hospital, or morgue any time of the day or night. In larger funeral homes, funeral directors might do shift work, including late nights, evenings, and weekends. In smaller funeral homes, working hours vary, but in larger homes, employees usually work eight hours a day, five or six days a week.

The job of a funeral director requires tact, discretion, composure, and compassion. When dealing with grieving people, a funeral director should be sympathetic, calm, patient, supportive, and kind. A funeral director provides support to the bereaved during the initial stages of their grief. He or she should have the desire and ability to comfort people in their time of sorrow and make their pain more tolerable.

To show proper respect and consideration for the families and the dead, funeral directors must dress appropriately. The profession usually requires short, neat haircuts and trim beards, if any, for men. Suits and ties for men and dresses for women are customary for a conservative look.

Funeral directors strive to foster a cooperative spirit and friendly attitude among employees and a compassionate demeanor toward the families. A growing number of funeral directors also organize post-death support groups and refer survivors to grief counselors to help them adapt to changes in their lives following a death.

Paperwork

Funeral directors are responsible for the success and the profitability of the businesses they run, and that means they must do a lot of paperwork. Directors

Some states now require funeral homes to take DNA samples, adding to the load of paperwork with which funeral directors must be concerned.

must keep records of expenses, purchases, and payments for their services. They have to submit reports for unemployment insurance; prepare federal, state, and local tax forms; and draw up itemized bills for customers. They also write obituary notices and have them placed in newspapers. Funeral directors increasingly use computers for billing, bookkeeping, and marketing. Some are beginning to use the Internet to communicate with clients who are preplanning their own funerals, or to assist family members who are developing electronic obituaries and guest books.

A funeral director also helps the deceased's family members apply for veterans' burial benefits and notifies the Social Security Administration of the death. Also, funeral directors may apply for the transfer of any pensions, insurance policies, or payments due on behalf of survivors.

Most funeral homes are small, family-run businesses, and the funeral directors are either owner-operators or employees. Most funeral homes have a chapel, one or more viewing rooms, a casket-selection room, and a preparation room. An increasing number also have a crematory on the premises. Equipment may include a hearse, a flower car, limousines, and sometimes an ambulance. Funeral homes usually stock a selection of caskets and urns for families to purchase.

After the Funeral

Funeral directors must be familiar with the funeral and burial customs of many faiths, ethnic groups, and fraternal organizations. The United States and Canada have been populated by immigrants from all over the world. Funeral practices and rites vary greatly among various cultures and religions. Some faiths have viewing; others do not. Some religions allow or even insist upon cremation, while others prohibit it. Some religions do not allow the bodies of the deceased to be embalmed.

Most commonly in the United States and Canada, the deceased are buried or cremated. A very small number choose to be cryogenically frozen. Among the many religions, nationalities, and ethnic groups in the United States, funeral practices usually share some elements.

Burial

After the funeral service comes the burial. Some people opt for graveside memorial services, but

The body of former Guyanese president Cheddi Jagan is cremated during a public ceremony in the town of Port Mourant, Guyana, on March 12, 1997. Most cremations in the United States and Canada are done privately in special furnaces.

often services are held in a place of worship or in a chapel at the funeral home. At a graveside service, the casket is taken to the cemetery and lowered into a hole dug by the cemetery groundskeeper or gravedigger before the funeral party arrives.

Family, friends, political leaders, and constituents pay their last respects to former North Carolina governor Terry Sanford during a funeral service in the Duke Chapel in Durham, North Carolina, on April 22, 1998.

As memorials vary between customs and faiths, so do burial rites. Direct burial is the norm for some cultures, such as Orthodox Jews. The body is taken from the place of death directly to the cemetery and buried in a simple container. If desired, a memorial service may be held later.

Cremation

Cremation is the burning of the deceased's body in a special furnace. People are increasingly choosing to be cremated after they die because it can be more convenient and less costly. Some environmentalists think it is better for the environment than filling up miles of scarce land with cemeteries. Cremation offers a range of options for final disposition. Cremated remains—or "cremains" in funeral industry lingo—can be retained in an urn or disposed of in a more elaborate manner. Some people elect to have their ashes dropped from an airplane, and there is even a program that, for a considerable sum, will send a small portion of the cremains into outer space!

The key thing to remember is that cremains are not ashes at all. They are the remains of the larger bones, which are then pulverized into a consistency somewhere between sand and kitty litter. They can be shipped via any traceable means, including the U.S. Postal Service, UPS, and Federal Express. For a death away from home, cremation is a way of transporting the remains far less expensively than transporting a body, and with less paperwork.

With cremation, memorial services can be held anywhere and at any time, sometimes months later, when all relatives and friends can get together. Even when remains are cremated, many people still want a funeral service. A funeral service followed by cremation need not be any different from a funeral service followed by a burial. Usually, cremated remains are placed in some type of permanent receptacle, such as an urn, before being committed to a final resting place. The urn may be buried, placed in the home, or housed in a special urn garden that many cemeteries provide for cremated remains.

In some cultures, entombment in a tomb or a mausoleum is very popular. The ashes or body of the deceased is laid to rest in a sealed vault that may be in an indoor or outdoor tomb or inside a mausoleum building. There are job opportunities in tomb guarding and in the care and maintenance of tombs and mausoleums. If you are interested in learning more about these jobs, please explore the funeral industry Web sites listed in the For More Information section at the back of this book.

3

The Funeral Director: Scientist and Artist

The career of the funeral director is a combination of jobs and is never boring. The funeral service professional must have a scientific education to handle dead bodies, artistic talent to improve their appearances, and caregiving skills to help survivors. You'll find that far from being morbid, funeral directors help both the living and the dead. They help the survivors come to terms as gently as possible, while sending off the dead with dignity.

Embalming

Almost all funeral directors also are trained, licensed, and practicing embalmers. Embalming is the art of preserving bodies after death. Technically, it is the removal of blood and replacement with formaldehyde-based fluid (formaldehyde is a chemical that preserves and disinfects formerly living things, from whole animals to organs and tissues).

Embalming is the most technical part of the funeral director's job. Embalmment is believed to have originated among the Egyptians, probably before 4000 BC, and was used by them for more than thirty centuries. Much evidence demonstrates that it is religious in origin, conceived as a means of preparing the dead for life after death. From the Egyptians, the practice of embalming spread.

Ancient embalming methods involved removing the brains and viscera (internal organs of the body), and filling the body cavities with a mixture of balsamic herbs and other substances.

Dutch anatomist Fredrik Ruysch is believed to have done the first modern embalming by injecting a prepared preservative chemical solution into blood vessels. During the nineteenth century, French and Italian scientists perfected such techniques, thereby enabling them to reach every part of the cadaver.

Embalming did not become popular in the United States until the Civil War, when slain soldiers needed to be shipped to their homes all over the United States. Before then, bodies were kept on ice and buried within a few days of death, but during the war, shipping often took several days by rail or horseback. Dead bodies begin to decompose immediately, often causing or spreading disease. Thus a way to preserve the bodies and keep them from spoiling was needed before they reached home for a proper funeral. Most corpses in the United States and Canada are embalmed, and the practice is widespread in other countries.

Modern embalming is important for preserving bodies until the funeral and to prevent the spread of infection both before and after burial. Embalming disinfects and preserves the remains of a person who has died of a disease. Contagious diseases have killed millions of people throughout the world and can spread quickly in unsanitary conditions. Embalming reduces the risk of people catching these diseases. Funeral directors and others who come into contact with diseased remains are at a very high risk of contracting the disease. Disinfection is important for funeral directors and for the safety of their communities.

Embalming is also done to slow down decomposition. Decomposition is the natural process of the body breaking down by bacteria. Decomposition requires only two elements: air and water, which make ideal hosts for bacteria. The human body begins to decompose as soon as death occurs. If more than twenty-four hours elapse between death and burial, state laws usually require that the remains be refrigerated or embalmed. Often, several days pass between a person's death and his or her funeral. Funerals are often delayed due to criminal investigations and autopsies, and until all of the family members and friends can assemble.

Without embalming, most remains would become unfit for public viewing within a very short time. A dead body undergoes constant chemical and physical changes that alter the appearance of the deceased. Embalming helps slow down this process and gives the living time to pay respect to the deceased.

The Procedure Requires Scientific Knowledge

Embalming involves many different steps and procedures. In fact, entire books are written on the embalming process. It is explained here only to give you a general idea of what must be done; you will learn more about embalming if you choose to pursue a career as a funeral director.

When a body arrives at a funeral home, the remains are laid out on a stainless steel or porcelain embalming table. All of the clothing is removed and either cleaned and returned to the next of kin or, if it is damaged or carrying disease, destroyed. Next, an inventory of any jewelry on the body must be carefully taken; rings must be taped or tied in place so they do not disappear. Other jewelry and glasses that are removed during embalming may then be replaced on the remains for the viewing.

The body is then cleaned by sponging it with disinfecting solution or spraying it with disinfectant spray. Next, the remains must be positioned. The legs of the corpse must be flexed, bent, and massaged to relieve rigor mortis, which is the stiffening of muscle tissue due to decomposition. The limbs are then moved to a presentable position, usually with the legs extended and arms at the sides.

With a scalpel or other medical instrument, a small incision is made on the right side of the lower neck. This is the location of the largest blood vessels in the body: the carotid artery and the jugular vein. Incisions are made in both vessels,

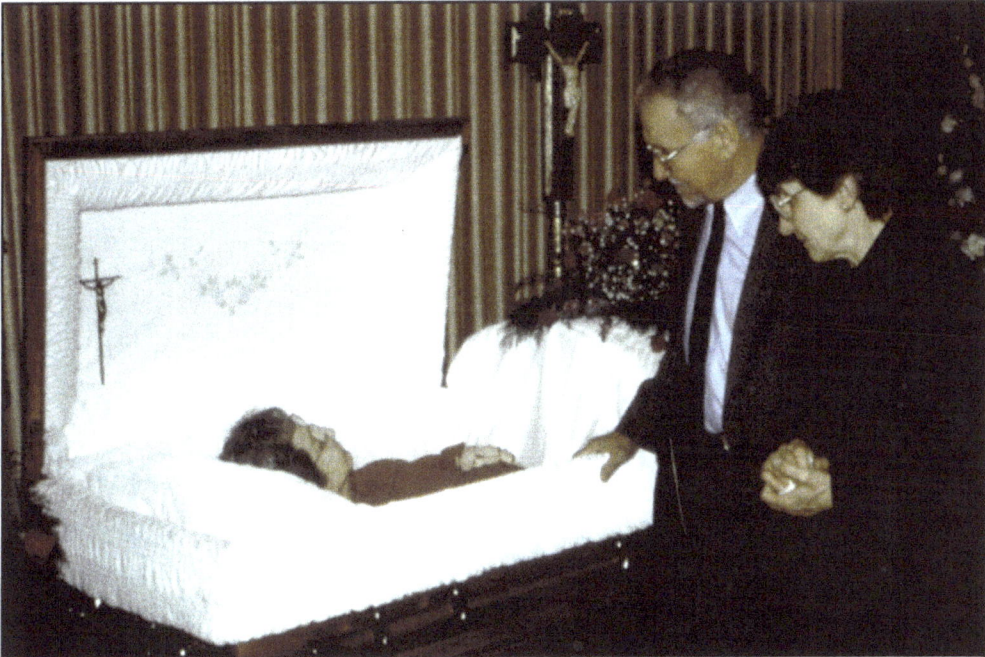
Adult siblings grieve the loss of their mother at a wake. Without embalming to slow down a corpse's decaying process, scenes like this would not be commonplace.

and a tube connected to an embalming fluid pump is inserted into the carotid artery. Another tube, called a drain tube, is put into the jugular vein. These tubes are attached to a machine that pumps embalming fluid into the artery, forcing the blood through the veins and out for disposal. The machine pumps about three gallons of a mixture of fluid and water through the remains to ensure they are thoroughly disinfected and preserved.

In most cases, this will be the only point of injection of the embalming fluid. There are times, however, when blood clots stop the flow of fluid through the whole system, and other points of injection must be found in order to do a complete and thorough embalming. This is not a career for the squeamish.

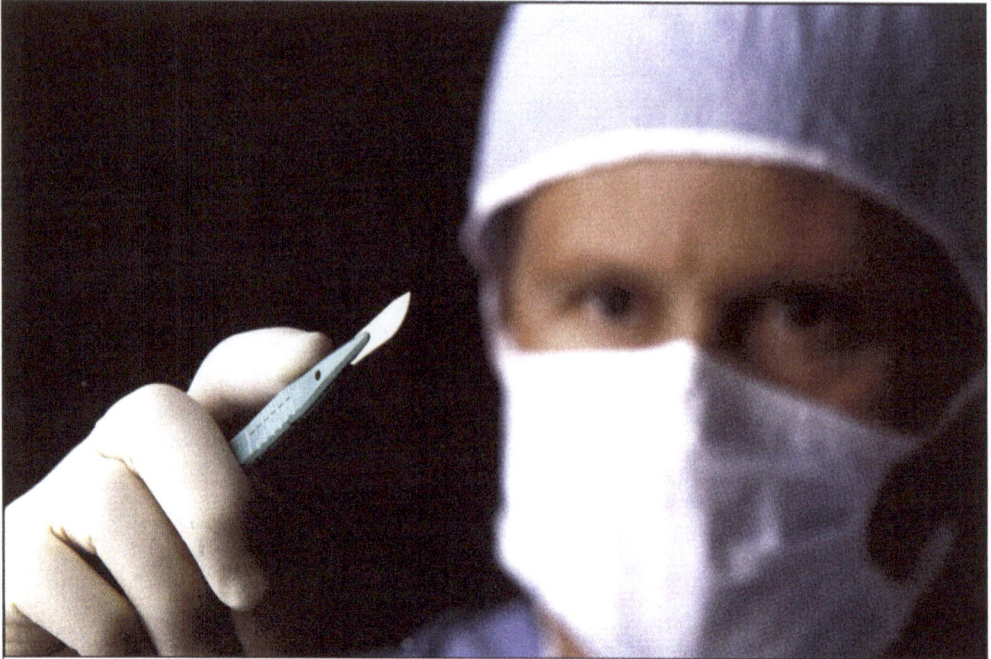

A funeral director prepares to embalm a body. He will make small incisions in the neck in order to replace the body fluids with chemicals.

The next step in the process is called cavity embalming. A small cut is made just above the navel and a suction pump is used to remove the blood and other bodily fluids. Next, a full-strength fluid is applied to the internal organs of the body, and all the incisions are then sewn up.

If you pursue an education in this field, you will learn more technical procedures in the process of embalming.

The Funeral Director as Artist

After the inside of the body is taken care of, the outside must be tended to. This is the point when the personal quality of compassion comes into the process. A funeral director must care for the deceased in such a way that will ease the suffering and grief of his or her family and friends.

First, the body is washed with cool water and a soapy, germicidal solution containing bleach to kill viruses and bacteria. The funeral director cleans the fingernails, shaves any facial stubble, applies chemicals to prevent scaling on the hands and face, and washes the hair, if needed. (A hairdresser is normally employed to style the hair after embalming is done.)

The next step is dressing and "casketing" the remains. The body is dressed, usually in clothes provided by the family. It is common to use a full set of clothing, including underwear, socks or stockings, and sometimes even shoes, if so desired.

Then makeup is applied to the face and hands of the body to make the person look as pleasant as possible. Usually, a special mortuary cosmetic is used, although store-bought cosmetics may be applied. Cosmetic work is the true art of the funeral director. It is through the proper application of cosmetics that a more lifelike presentation will be made. It helps the survivors to see their loved one as he or she looked when alive. Using too little or too much makeup affects the appearance of the body. The coloring of the deceased before death must be determined (often by looking at photographs of him or her), and the makeup adjusted accordingly.

The final step is placing the remains in a casket, properly adjusting the interior of the casket, and then touching up the clothing, hair, and cosmetics. This final step takes precision and

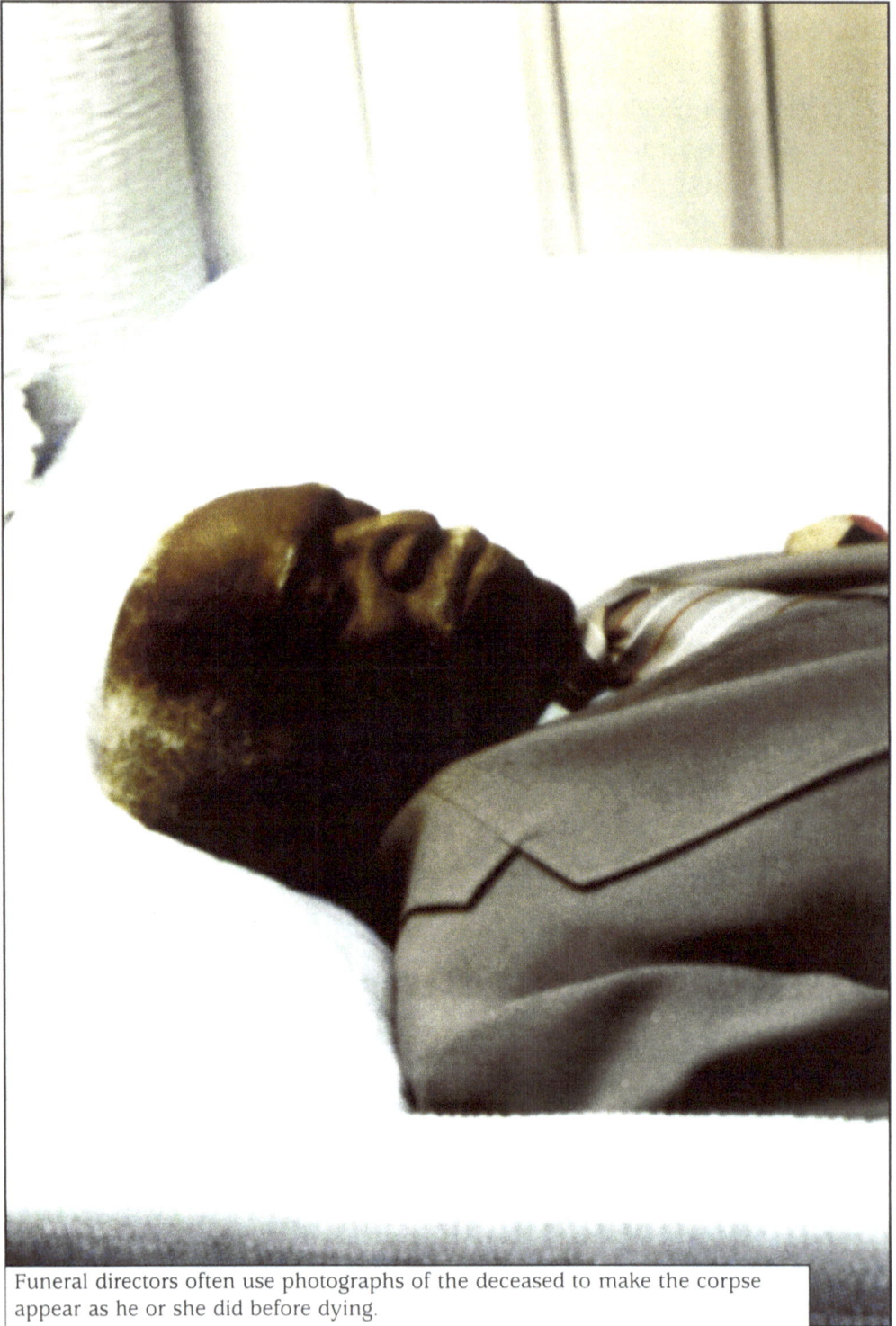

Funeral directors often use photographs of the deceased to make the corpse appear as he or she did before dying.

is usually very time consuming. The funeral director tries to pose the head and hands in a lifelike position and finishes up by making everything look impeccable.

4

Preparing for a Career in the Funeral Industry

If you choose to become a medical examiner, coroner, or pathologist, you must work hard and get a good education to prepare for the job. Working in the death industry is challenging; you will see many unpleasant things. Many people find these careers rewarding because they get to use their knowledge to solve mysteries. As a coroner or medical examiner, you will have to use science to find the answers to many questions.

A funeral director needs a combination of training and skills: the education of a scientist, the talent of an artist, and the compassion of a therapist.

How to Become a Coroner or Medical Examiner

To become a medical examiner, you will need to go to medical school and become a medical doctor. What you study to become a doctor would fill a book bigger than this one. Likewise, if you are interested in a career as a coroner, you generally

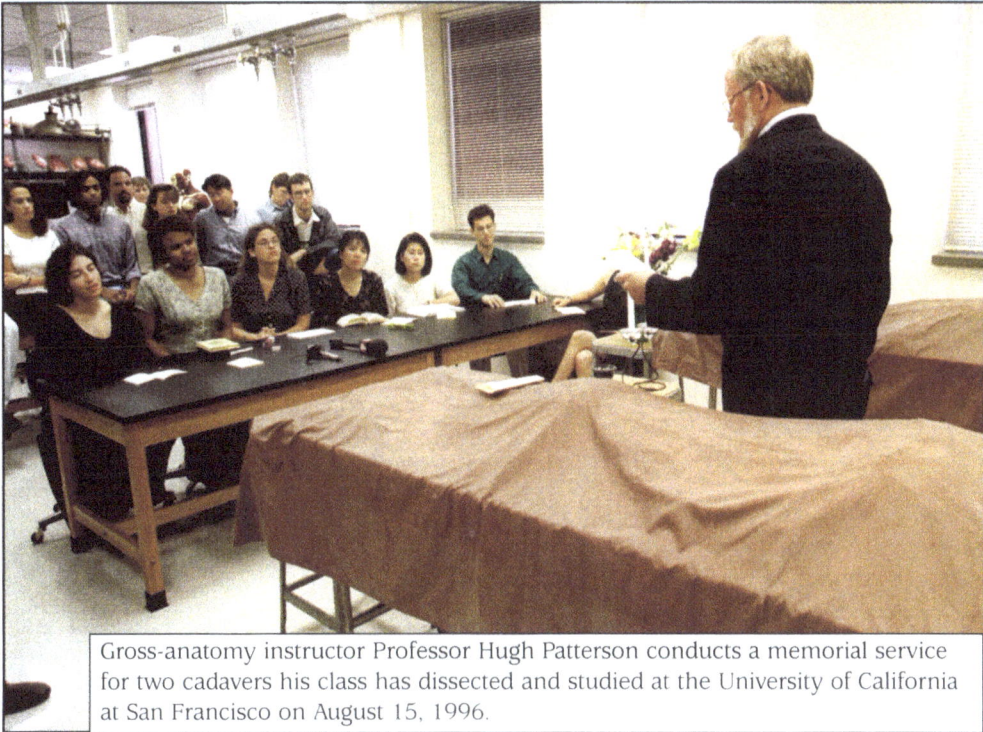
Gross-anatomy instructor Professor Hugh Patterson conducts a memorial service for two cadavers his class has dissected and studied at the University of California at San Francisco on August 15, 1996.

need a medical degree, but the laws vary, as noted in chapter 1.

Medical examiner and coroner programs include courses in anatomy, physiology, pathology, embalming techniques, restorative art, business management, accounting, and use of computers in funeral home management and client services. They also include courses in the social sciences and legal, ethical, and regulatory subjects, such as psychology, grief counseling, oral and written communication, funeral service law, business law, and ethics.

After you have completed your formal training, you generally must serve a residency period and begin your career in a junior-level or assistant position. Medical examiners and coroners are often appointed or elected in the region where they live.

Becoming a Pathologist

After high school, you will attend college for four years to earn a bachelor's degree. Next will be four years in medical school, until you earn an M.D. degree. After medical school, there are several routes you can choose to become a forensic pathologist. You may spend five years training in anatomic and clinical pathology, followed by one year of residency or fellowship in forensic pathology. A second option is to train for four years in anatomic pathology, and then for one year in forensic pathology. The residency training in forensic pathology involves practical, on-the-job experience supervised by trained forensic pathologists. Residents actually perform autopsies and participate in death investigations. After your training, you must pass an examination given by the American Board of Pathology to become certified. If you pass the test, you are considered board certified and can legally practice.

The road to becoming a pathologist is long, but the career can be lucrative. Salaries in forensic pathology start around $40,000 and go to well over $100,000 with years of experience. Forensic pathologists work for states, counties, groups of counties, or cities, as well as for medical schools, the military, and the government. In medium-sized and smaller areas, the forensic pathologist may work for a private group or hospital that contracts with the county to perform forensic autopsies. Although some of the work may seem unpleasant, pathologists practice in the finest tradition of

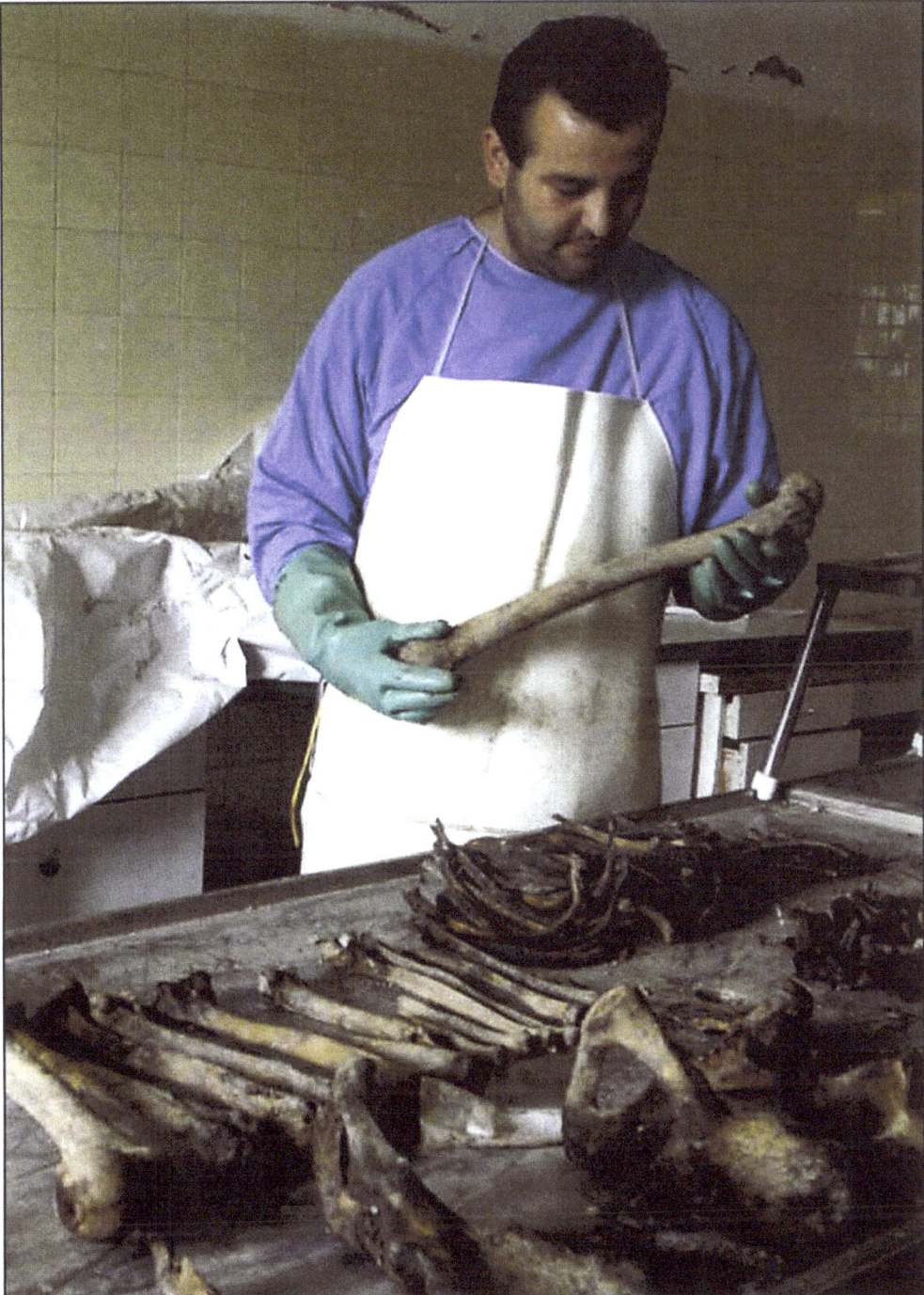

A pathologist examines human remains found in a mass grave in Bosnia. His findings support war crimes investigations against former Serbian president Slobodan Milosevic.

preventive medicine and public health by making the study of the dead help the living.

Other Careers in Forensics

In addition to the established specialties discussed earlier, there are many new areas of forensic study that are growing in importance. Artists and sculptors use their talents to create reconstructions that identify assailants and remains that are found. Computers are playing an ever increasing role in the forensic sciences: Forensic computer-related crime investigators and forensic image enhancement specialists rely on them. Enhancement specialists attempt to gather vital evidence such as fingerprints from exhibits using highly technical methods.

There also are specialists who work in the fields of forensic education, forensic administration, forensic research, and forensic rehabilitation. Many of the professionals in the various disciplines wear different hats in terms of their fields of interest. They may be involved in forensic laboratory investigation, field investigation, clinical work, communications, education, administration, and research.

Forensic nurses assist pathologists at crime scene investigations and in medical settings like hospitals and rape crisis centers. The gathering of evidence and the conducting of investigations may be done by forensic coroners, forensic crime scene investigators, or forensic death investigators. Other specialists who uncover information are forensic photographers, forensic polygraph (lie-detector test) examiners, forensic radiologists,

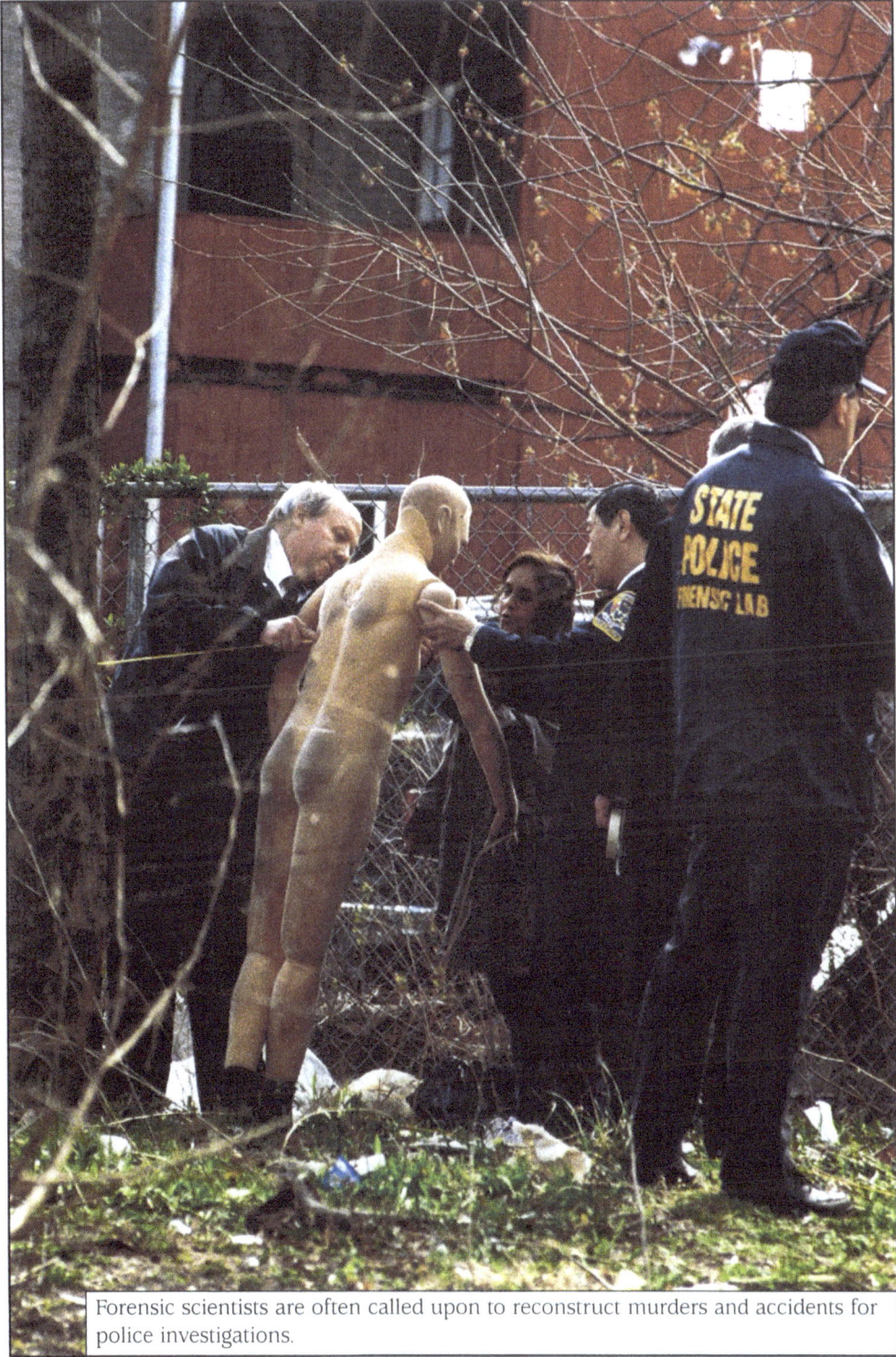

Forensic scientists are often called upon to reconstruct murders and accidents for police investigations.

and forensic speech scientists, who enhance recordings and examine the authenticity of transcripts and recordings.

Pursuing a Career as a Funeral Director

According to the Bureau of Labor Statistics for the U.S. Department of Labor, job opportunities for funeral directors are best in larger funeral homes— funeral directors may earn promotions to higher-paying positions such as branch manager or general manager. In 1998, almost one in ten were self-employed, meaning they own the funeral home where they practice their craft.

Some directors eventually acquire enough money and experience to establish their own businesses. Salaries vary depending on the size of the area in which the director works. Funeral directors in large cities earn more than those in small towns and rural areas. According to the U.S. Department of Labor, funeral directors earned an average income of $35,000 in 1999. The highest paid funeral directors made more than $78,500 per year; the lowest salary was $17,000. A funeral director's salary depends on his or her number of years in funeral service, services performed, the number of facilities the company operates, the area of the country and the size of the community where he or she lives, and the level of education he or she has obtained.

A high school student can start preparing for a career as a funeral director by taking a lot of

Showing families how to personalize a funeral with memories that reflect the deceased is a growing trend among funeral directors.

science courses, especially biology and chemistry. You should learn all you can about the human body—you're going to be dealing with it every day. Also, take some business and communications courses. Summer jobs or part-time after-school internships are a great way to jump-start any career. Be warned: Many part-time jobs in funeral homes involve maintenance and clean-up tasks, such as washing and polishing limousines and hearses. That's part of the business. One female funeral director admitted to mowing part of the cemetery lawn just before the burial while wearing a business suit and high-heels! You've got to be prepared to do whatever is necessary to make a funeral run smoothly.

You should strive to go to an accredited school that specializes in mortuary science. College programs in mortuary science usually last from two to four years. The American Board of Funeral Service Education accredits forty-nine mortuary science programs in the United States. Two-year programs are offered by a small number of community and junior colleges, and several colleges and universities offer both two- and four-year programs.

In college, you can expect to take a lot of science courses, such as chemistry, biology, microbiology, anatomy, and pathology. You will also have extensive courses in restorative art and embalming. Since you will be dealing with people who are very upset, it helps to know a little psychology. Courses in social sciences may be part of your education, with classes about the psychology of grief, funeral directing, the history of funeral service, and counseling skills.

The funeral service industry is highly regulated by the state and federal governments, so you will have to know about government regulations and some legal issues. You may take classes in business law and funeral service law, and regulation and professional ethics. Knowing how to manage a business is also important. You will take courses in business management, funeral arranging, funeral merchandising, funeral home management, computer applications, Federal Trade Commission funeral rules, and accounting.

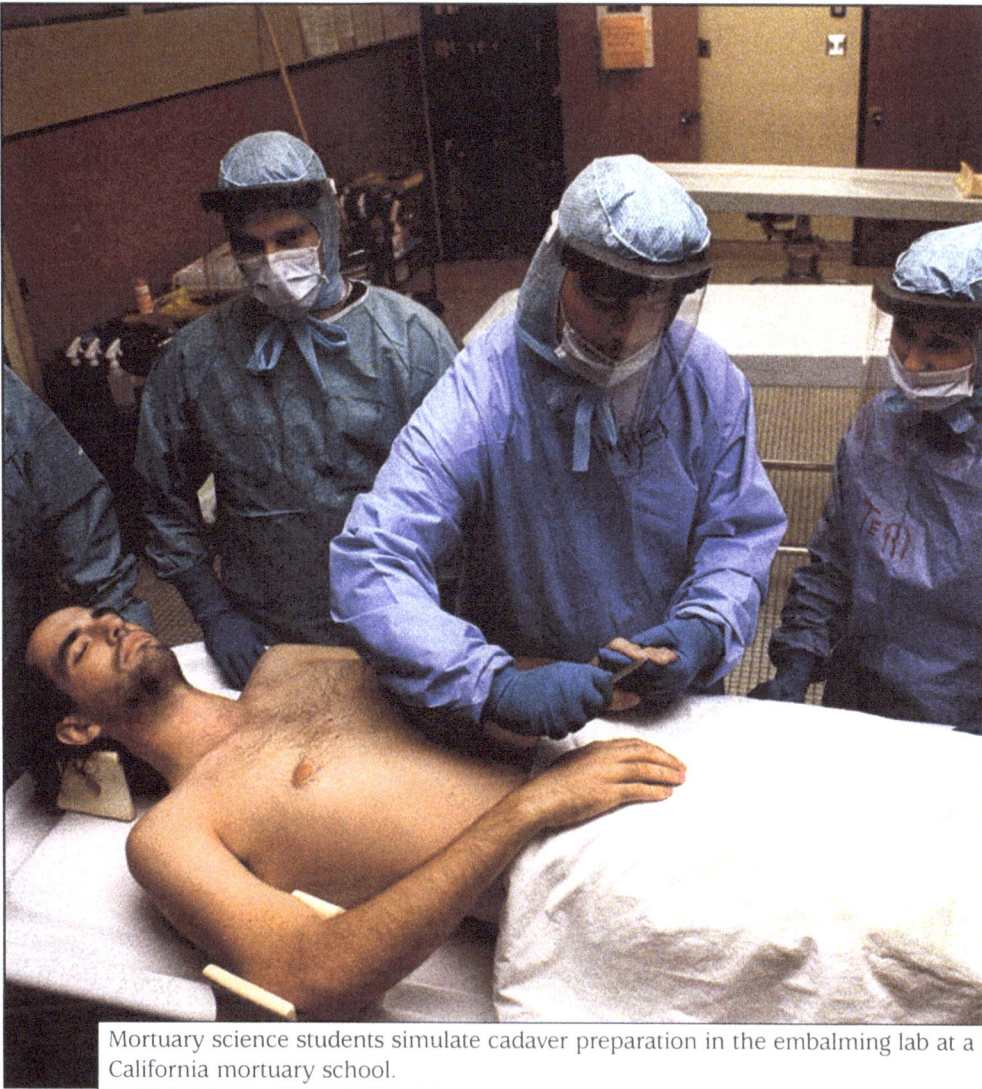

Mortuary science students simulate cadaver preparation in the embalming lab at a California mortuary school.

Generally, you have to serve an internship or work as an apprentice under an experienced and licensed funeral director or embalmer. Depending on the laws in your area, apprenticeships last from one to three years and may be served before, during, or after mortuary school. Apprenticeships provide practical experience in all areas of the funeral service profession, from embalming to digging graves.

Many funeral homes use computerized catalogs to enable the bereaved to choose elements of a funeral without having to physically visit a showroom.

Continuing education is important. Part of working in a technical profession such as the funeral industry is staying up-to-date on new advances, technology, and research. Technology is improving how everything is done, and keeping up with changes is an important responsibility in the funeral industry.

In the United States, funeral directors and embalmers must be licensed (except in Colorado). Licensing laws vary from state to state, but most require applicants to be at least twenty-one years old, have two years of formal education that includes studies in mortuary science, serve an apprenticeship, and pass a qualifying examination. After becoming licensed, new funeral directors may join the staff of a funeral home.

Anyone who will be performing embalming procedures must be licensed in all states, and some states issue a single license for both funeral directors and embalmers. In states that have separate licensing requirements for the two positions, most people in the field obtain both licenses.

Funeral directors occasionally come into contact with the remains of persons who had contagious diseases, but the possibility of infection is remote if strict health regulations are followed. They must know all of the current rules and regulations on corpse handling.

Other Careers in the Funeral Industry

If college is not in your future, there are other careers in the funeral industry that you can pursue.

Funeral homes usually stock a wide array of caskets to meet the varied preferences of the surviving families and, in the case of prearranged funerals, the deceased.

Most involve working for funeral homes. You could make and sell caskets and vaults, or work in a cemetery, digging graves. You could work in the crematorium. You could carve monuments and make grave markers. (This usually requires some artistic ability and an apprenticeship in stonecutting.) There are also jobs available in pet-related death care. Every big city has at least one company that specializes in trauma scene cleanup. You could even be a hearse driver with a professional car rental agency or livery service.

A growing number of companies will arrange a person's funeral while he or she is still living. You do not have to be a funeral director to put together all of the details. People who prearrange funerals contract services from other funeral service providers, and clients pay in advance. Prearranging funerals is increasing in popularity because it can provide peace of mind by ensuring that the client's wishes will be taken care of in a way that is satisfying to him or her and to those who will survive.

The National Funeral Directors Association also predicts an increase in funeral home "aftercare" services, including support groups, remembrance services, and community referrals. Immigration trends mean that funeral directors throughout the United States and Canada now serve people with funeral customs from all parts of the globe.

5

The Future of the Funeral Industry

According to the U.S. Department of Health and Human Services, there were 2,345,702 deaths in the United States during 1999. Of these deaths, 75 percent resulted in earth burial or entombment, and 25 percent resulted in cremation. The Centers for Disease Control in Atlanta, Georgia, reports that autopsies are performed in 14 percent of all deaths, but that number jumps to 97 percent in instances of homicide.

According to the National Funeral Directors Association, there are more than 22,100 funeral homes in the United States. These employ approximately 35,000 licensed funeral directors and embalmers, and 89,000 additional funeral service and crematory support personnel. In addition, opportunities are increasing within the industry. The number of women and minorities entering the funeral service profession is growing. In fact, according to the NFDA, approximately one-third of current mortuary-science enrollees are women.

A mourner scatters dirt over the casket of a loved one during a graveside service.

The need to replace funeral directors and morticians who retire or leave the occupation for other reasons will account for even more job openings than employment growth, according to the Department of Labor (DOL) Occupational Outlook Handbook. Compared to workers in other occupations, reports the DOL, more funeral directors have reached the age of fifty-five or older, and will be retiring in greater numbers through 2008. Although employment opportunities for funeral directors are expected to be good, mortuary-science graduates may have to relocate to find jobs in funeral service.

More people are choosing careers in the funeral industry. In fact, the DOL estimates that employment opportunities for funeral directors should increase by 10 to 20 percent through 2008. The proportion of elderly people is projected to grow during the coming decade. Consequently, the number of deaths is expected to increase, spurring an even greater demand for death investigations and funeral services.

Funeral industry professionals provide care for the dead and comfort for the living. If you are a person who has a strong interest in helping people, one of the many careers in the funeral industry could be right for you.

Glossary

anatomy The study of the muscles, organs, glands, bones, and structure of the body.

autopsy The examination of the body, inside and out, to determine the cause of death and to identify and understand diseases.

casket A large chest or long box, often called a coffin, made of varying kinds of wood or metal, in which remains are buried.

cemetery A burial ground for the dead.

contagious Easily spread between people or animals.

coroner A public official who investigates deaths.

cremate To burn a corpse.

cryogenics The storing of deceased remains in very low temperatures.

custom A traditional act that is passed down through a family, a community, or a culture.

death certificate A document that states the official cause of death as determined by a medical doctor and verified by a public official.

decomposition The breaking down and decaying of a body by bacteria; the natural process that occurs in all living things immediately after death.

disposition The final settlement of something.

embalm Method of pumping chemicals into a body to slow down decomposition and prevent the spread of disease.

entombment To bury a dead person in a chamber, usually one that is above the ground.

formaldehyde A chemical substance used to preserve dead organic matter, such as organs.

funeral A ceremony held to commemorate, celebrate, and say good-bye to a friend or loved one before his or her burial or cremation.

funeral director The person who arranges and directs all of the activities involved with burying a dead person. Also called a mortician or undertaker.

infectious Something that spreads easily between people.

medical examiner An appointed public official who conducts autopsies to find out what killed a person.

morgue A sterile environment, usually in a hospital, where corpses are brought for autopsies.

mortuary science The study of the many aspects of the funeral industry.

pathologist A doctor trained in the branch of medicine that deals with the diagnosis of disease and causes of death.

pathology The field of medicine that studies the causes of disease and death.

rigor mortis The stiffening of muscles and joints due to decomposition.

survivor One who outlives a person who has died.

tomb Chamber or vault in which the deceased is laid to rest. Often used interchangeably with mausoleum.

toxins Poisons, either chemical or naturally occurring, that cause sickness and death.

visitation The time when friends and family gather in a funeral home to see the deceased prior to burial. Also known as a wake.

For More Information

In the United States

American Academy of Forensic Sciences
P.O. Box 669
Colorado Springs, CO 80901-0669
(719) 636-1100
Web site: http://www.aafs.org

American Board of Funeral Service Education
38 Florida Avenue
Portland, ME 04103
(207) 878-6530
Web site: http://www.abfse.org

Association for Death Education & Counseling
342 North Main Street
West Hartford, CT 06117-2507
(860) 586-7503
email: info@adec.org
Web site: http://www.adec.org

Cincinnati College of Mortuary Science
645 W. North Bend Road
Cincinnati, OH 45224
(513) 761-2020
e-mail: generalinfo@ccms.edu.
Web site: http://www.ccms.edu

International Cemetery and Funeral Association
1895 Preston White Drive, Suite 220
Reston, VA 20191
(703) 391-8400
(800) 645-7700
email: gen4@icfa.org
Web site: http://www.icfa.org

National Association of Medical
   Examiners (NAME)
1402 South Grand Boulevard
St. Louis, MO 63104
(314) 577-8298
e-mail: name@slu.edu
Web site: http://www.thename.org

National Funeral Directors Association (NFDA)
13625 Bishop's Drive
Brookfield, WI 53005
(800) 228-6332
e-mail: nfda@nfda.org
Web site: http://www.nfda.org

In Canada

Canadian Association of Pathologists
774 Echo Drive
Ottawa, ON K1S 5N8
(613) 730-6230
(800) 668-3740 ext.230
email: cap@rcpsc.edu
Web site: http://cap.medical.org

Canadian Society of Forensic Science
2660 Southvale Crescent, Suite 215
Ottawa, ON K1B 4W5
(613) 738-0001
Web site: http://www.csfs.ca/greet.htm

The Funeral Service Association of Canada
Harwood Avenue South, Suite 201-206
Ajax, ON L1S 2H6
(905) 619-0982
email: info@fsac.ca
Web site: http://www.fsac.ca

The Funeral Services Information Portal and
    Directory
P.O. Box 2195
Richmond Hill, ON L4E 1A4
(905) 773-1767
email: nicole@thefuneraldirectory.com
Web site: http://www.thefuneraldirectory.com/

Web Sites

Careers and Education in Forensics
http://home.istar.ca/ ~ csfs/caremp.htm
Good source of information regarding
   employment, careers, and education in the
   field of forensics.

Forensic Science Links and Information
http://dying.about.com/health/dying/cs/
   forensicscience1/index.htm

Forensic Science Web Page
http://home.earthlink.net/ ~ thekeither/Forensic/
   forsone.htm

Funeral Service Educational Foundation
http://www.fsef.org
e-mail: fsef@fsef.org

National Funeral Directors & Morticians
http://www.nfdma.com

UCLA Pathology Department
http://www.bol.ucla.edu/ ~ pathres/sites/autopsy.htm
Useful pathology links

For Further Reading

Cronin, Xavier A. *Grave Exodus: Tending to Our Dead in the Twenty-First Century.* New York: Barricade Books, 1996.

Hatch, Robert T. *What Happens When You Die: From Your Last Breath to the First Spadeful.* Secaucus, NJ: Carol Publishing Group, 1995.

Iserson, Kenneth V. *Death to Dust: What Happens to Dead Bodies?* 2nd ed. Tucson, AZ: Galen Press, 2001.

Lynch, Thomas. *The Undertaking: Life Studies from the Dismal Trade.* New York: Penguin, 1998.

Mayer, Robert G. *Embalming: History, Theory & Practice.* 3rd ed. New York: McGraw-Hill, 2000.

Mitford, Jessica. *The American Way of Death Revisited.* New York: Vintage Books, 2000.

Sacks, Terence J. *Opportunities in Funeral Services Careers.* Lincolnwood, IL: VGM Career Horizons, 1997.

Smith, Ronald G. E. *The Death Care Industries in the United States.* Jefferson, NC: McFarland & Company, 1996.

Index

About the Author

Nancy L. Stair is a writer and editor who lives in Greenwich Village, New York City, with her sweetheart Charley, a professional trombone player.

Photo Credits

Cover, p. 41 © Corbis; pp. 2, 7, 10, 12, 15, 23, 25, 26, 39, 43, 45, 47, 48, 50, 53 © AP/World Wide Photos; p. 34 © Index Stock; pp. 33, 36 © The Image Works.

Design

Geri Giordano

Layout

Nelson Sá

www.ingramcontent.com/pod-product-compliance
Lightning Source LLC
Chambersburg PA
CBHW050910210326
41597CB00002B/80